The Beachcomber's Guide to Fossils in Capitola

Dear Ava, and so are you ♡

I'm the best. For real

thanks so much for

coming, love you nerd,

Lorene Paras

The Beachcomber's Guide to Fossils in Capitola

LORENA PARAS

GRIZZLY PEAK PRESS
Kensington, California

For information contact:
Grizzly Peak Press
350 Berkeley Park Boulevard
Kensington, California 94707
grizzlypeakpress.com

ISBN Number: 978-0-9839264-7-4
Printed in the United States of America

Cover Photo courtesy of Maire O'Sullivan
Cover art and all other photos by the author.

Dedication

Dedicated to all my friends who didn't call me a huge nerd as often as they could have during this process.

Acknowledgments

First off, I'd like to thank my parents Barbara Englis and Melinda Paras, for being the official chauffeurs of the many trips to and from Capitola, in addition to paying my way through bookmaking class. I would also like to thank my friends who supported me and helped me make up class work I missed because of frequent trips down the coast or art classes on Fourth Street and my teachers for letting me skip school during those times. Which brings me to thanking Alice Armstrong and Castle in the Air for helping me make the original first copy of my book and helping me with the artistic elements. I also want to say thanks to Frank Perry, who wrote a book without which I would not have been able to identify some of the fossils I found, in addition to helping edit both the historical and scientific work. That leads me to thanking Jean DeMouthe and Chrissy Garcia from the California Academy of Sciences for helping me fact check and identify the rest of my fossils. Also thank you to Don Ellis for encouraging me to move this book to publication and his expertise in guiding the publication effort. I'd also like to express my gratitude toward Zac Rymland and Jeffrey Reynolds for their contribution of layout, typesetting, and design. And last but not least, I'd like to thank my Auntie Chela for the inspiration of my interest in the sea, her guidance in the publication of scientific work, and her living example of the beauty of art in nature.

Foreword

I am pleased to introduce this lovely book, which will provide the reader with a guide to some of the fossils that can be found at low tide along the cliffs at Capitola Beach in Santa Cruz County, California. The fossils are illustrated with photographs and drawings, and information about the animals, or their modern-day counterparts is provided, along with a brief history of Capitola.

The original version of this book was a handbound volume, which the author prepared for an 8th grade master works project for Black Pine Circle in Berkeley in 2012. As such, the book reflected the author's longtime interest in writing, marine biology, paleontology, art, and photography. (Indeed, Lorena's interest in marine biology had already been long-term by 8th grade: she had learned the Hawaiian names of fish by the age of 2, had assisted with data collection on hammerhead shark feeding at 5, and had photographed her first moray eel while snorkeling at age 8.) To produce the book, Lorena made numerous trips to Capitola and to the Santa Cruz Museum of Natural History, took a bookbinding class, learned the stippled ink drawing technique, and corresponded via e-mail with Frank Perry whose book *Fossil Invertebrates and Geology of the Marine Cliffs at Capitola, California* was a major inspiration for this one.

As she began high school, Lorena's friends and family suggested that she consider publishing the book. To do so, they realized that the book would need to meet a higher scientific standard, and asked me to review it. Among other things, I suggested that Lorena check her fossils against the collection of fossils from Capitola that are held at the California Academy of Sciences. Lorena did so with guidance from Jean DeMouthe, and subsequently revised species names and updated the book. The current version has been reorganized and professionally edited, but the content and writing otherwise remain the same as the original.

What you now hold in your hands is a field guide, but it is also a book about joy: the joy of time spent at the beach in the colorful town of Capitola-by-the-Sea, the joy of exploration and discovery, the joy of connecting oceans of the past to oceans of the present, and the joy of carefully and thoughtfully expressing that connection in writing and art. We hope it accompanies you to the beach and enhances the joy you find hunting for natural treasures on a beautiful summer day by the sea.

Chela Zabin, PhD
University of California Davis
Smithsonian Environmental Research Center
Santa Cruz, California
14 March 2015

Author's Note

Ever since I was little I have always loved the ocean. Cheap plane tickets and family friends on Oahu made Hawaii a semi-frequent vacation. My Auntie Chela would take me out tide pooling at the beach and I was fascinated by how many different organisms lived in such a small habitat. She'd help me pick up hermit crabs and teach me about the barnacles. After that I was sure I wanted to grow up to be a marine biologist like her. Of course after that I was also sure I wanted to be an archaeologist, a treasure hunter, and possibly a professional pirate. But somehow the marine biologist phase and subsequently the paleontologist phase both stuck.

Years later, in 8th grade at my middle school, Black Pine Circle, we were asked to come up with a project that we could potentially spend 40+ hours on. After my initial plan to make a somewhat sizable fireworks display was shot down by administration, I decided I wanted to do something to combine my interests. I settled on a project that would put together my knowledge of marine biology and paleontology with my enthusiasm for the arts. After working for definitely more hours than the school required, I came up with a shorter, less complete version of the book you are reading right now. Then, maybe half a year and an offer to assist in the publishing process later, I set out to improve my book into something that could imaginably be released out into the world.

Table of Contents

The Beachcomber's Guide to Fossils in Capitola

Beach houses

Capitola City Beach

Storefronts by the beach

Capitola Wharf and beach houses

Stockton Avenue Bridge

Capitola

Capitola, California, is a small beach town just east of Santa Cruz along Monterey Bay. It is known for being one of the oldest resorts on the California coast. It was built on the location of a Native American village that existed for more than a thousand years. The native inhabitants, known to the colonizers as the Soquel Indians, were removed to Mission Santa Cruz after it was established in 1791.

The original English name of Capitola was Soquel Landing because of a wharf at the mouth of Soquel Creek. In 1856, Frederick Hihn, after acquiring the land around the wharf, decided to turn the town into a seaside resort. He commissioned Sedgwick Lynch to build the iconic wharf in 1857. Then, to help development, Hinn signed a contract with a pioneer who built the nearby town of Soquel. His name was Samuel Hall, and he was a carpenter, craftsman, and farmer.

Camp Capitola's grand opening was in the summer of 1874. Hall had built a stable, cottages, bathhouses, and a hotel. Around 1900, Camp Capitola grew to Capitola by the Sea. A 160-room hotel called Hotel Capitola was built on the coast, designed by Edward Van Cleeck, but burned down in 1929. While standing, it was one of the largest hotels in the West.

The best-known story of how Capitola was named is that Samuel Hall's daughter named it after the heroine of the novel *The Hidden Hand* by E.D.E.N. Southworth.

From the outside, Capitola looks like a normal beach town. Small shops, restaurants along the water. A typical beach, surfers, swimmers, sunbathers. But if you head down the beach, there's something else. Along with the seagulls and algae, in the cliffs and the rocks, there's something much older. The fossils of the Purisima Formation.

Lorena Paras

Fossil layers in Capitola

Capitola Bluffs

Lorena Paras

The Purisima Formation

Millions of years ago, the ocean covered what is now Capitola. Many creatures lived in its depths and when they died their bones, shells, tests, or anything that remained would sink to the sandy bottom. Over time layers of sediment piled up. Later, when the sea level dropped and the water receded, the sediment from rivers and the bottom of the ocean remained, hardened into rock, forming the cliffs between Capitola and New Brighton beach. As parts of the cliffs fell away the fossils were revealed. They are part of the Purisima Formation.

The Purisima Formation is a rock formation going all the way from Point Reyes in the north to the Santa Cruz Mountains in the south. The cliffs vary from mudstone to medium-grained sandstone. Exposures of the Purisima occur in three different structural blocks—the Santa Cruz, Pigeon Point, and Point Reyes tectonic blocks. The formation takes its name from the Purisima Creek in San Mateo County. The fossils in the Purisima Formation can be anywhere from 3 million to 5 million years old.

The fossils found in the Purisima Formation are from the late Miocene and Pliocene ages. The Miocene extends from 23 million to 5 million years ago. Its name comes from the Greek words *μείων* (*meiōn*, "less") and *καινός* (*kainos*, "new") and means "less recent" because it has 18% fewer modern sea invertebrates than the Pliocene. The Pliocene age was much more brief. It went from about 5 million to 2 million years ago. The name comes from the Greek words *πλεῖον* (*pleion*, "more") and *καινός* (*kainos*, "new") and means roughly "continuation of the recent."

More fossils in the Capitola Bluffs

Shell fossils embedded in rock on Capitola Beach

Lorena Paras

Fossils of the Purisima Formation

There are many different fossils in the Purisima Formation, but they are mostly invertebrates. There are over 55 species of mollusks, six species of crustaceans, two species of echinoderms, brachiopods, and a multitude of microscopic invertebrate species. Mollusks are highly diverse, but 80% are gastropods (snails). Mollusks comprise about 23% of all (named) oceanic organisms. Crustaceans form a large group of arthropods, and include crayfish, lobsters, crabs, krill, shrimp, and barnacles. Echinoderms include sea urchins, sea cucumbers, and sea stars. Brachiopods are marine animals that have hard valves (shells) on the upper and lower surfaces, instead of the left and right arrangement in bivalve mollusks.

Because many of the fossilized species still live in and around Monterey Bay, we can envision what it might have been like millions of years ago. We know that these invertebrates probably lived at a depth of around 30 to 150 feet. The same depths occur today one to eight miles from shore in northern Monterey Bay.

To find these fossils all you need to do is head to Capitola Beach at low tide and walk around the concrete outcropping on the south end. Make sure you plan according to the tides, because you don't want to get stuck there during high tide.

Remnants of shell fossils in the cliffs

Individual Fossils

Bivalves

A class of freshwater and marine mollusks characterized by hinged shells.

Anadara trilineata

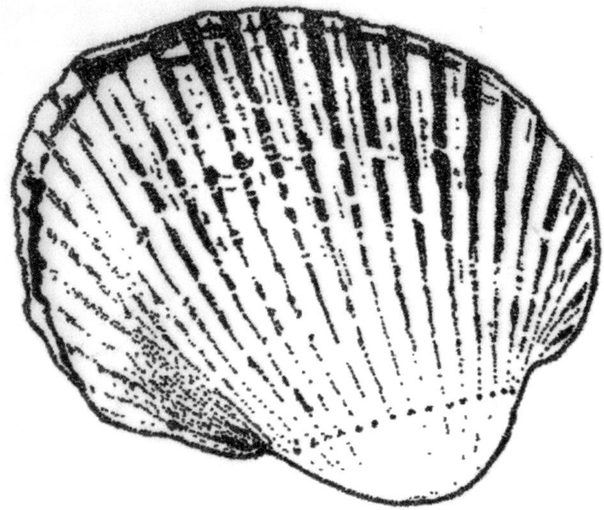

Lorena Paras

Ark Clam

Scientific Name: *Anadara trilineata*
Order: Arcoida
Family: Arcidae
Size Range: varied
Habitat: shallow water, buried in sediment

About: Ark clams are saltwater clams in the family Arcidae. There are over 200 living species of ark clams. Some species can live up to 10 years. Ark clams live in the sandy bottom of bays and inshore and offshore of the ocean. They have thick, sturdy shells to withstand strong currents. Red ark clams are considered a specialty in Japan and used in sushi and sashimi. They are called red ark clams because they are one of the only types of clams that have red blood. The "ark" part of the name possibly comes from the ridging on the shell that resembles many drawings of Noah's ark.

Mactromeris coosensis

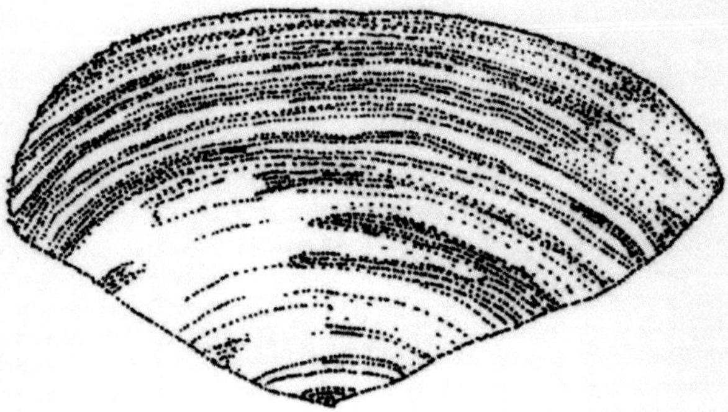

Surf Clam

Scientific Name: *Mactromeris coosensis*
Order: Veneroida
Family: Mactridae
Size Range: up to 5 cm
Habitat: offshore, low intertidal and surf zones

About: *Mactromeris coosensis* is a member of the family Mactridae, otherwise known as dish or surf clams. It is a bivalve mollusk that lives buried in the sand. A relative of Spisula coosensis, Spisula solidissima, lives on the East Coast of the United States and can grow up to 20 cm. It is commonly eaten as clam strips in sushi and chowder. Surf clams have been known to live past 30 years old. Like most clams, they are filter feeders, which means they use their siphon to suck in and filter plankton and other organic matter for food.

Clinocardium meekianum

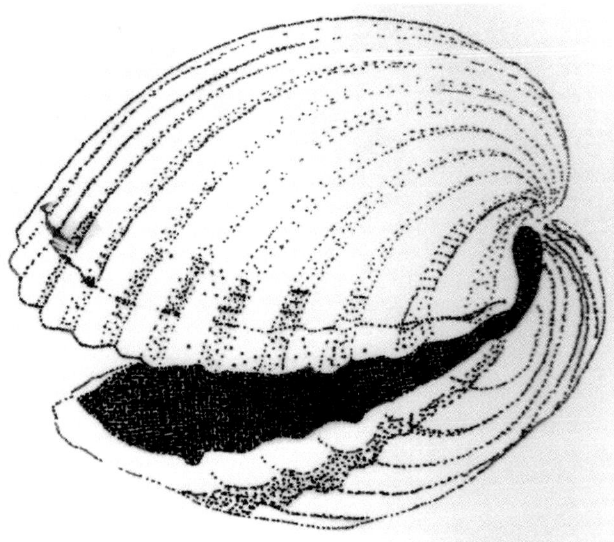

Cockle

Scientific Name: *Clinocardium meekianum*
Order: Veneroida
Family: Cardiidae
Size Range: 1–15 cm
Habitat: sandy, sheltered beaches

About: Cockles in the family Cardiidae are sometimes called "true cockles" due to the fact that shellfish vendors often mislabel other similar bivalves as cockles. Cockles burrow down in the sand using their foot and filter water for food. They mostly eat plankton. There are over 200 species of cockles, not including extinct species. Cockles are capable of jumping by bending and straightening their foot. They are edible, and popular in many cultures. In the United Kingdom, freshly cooked cockles are sold as a snack from seafood stalls.

Tresus pajaroanus

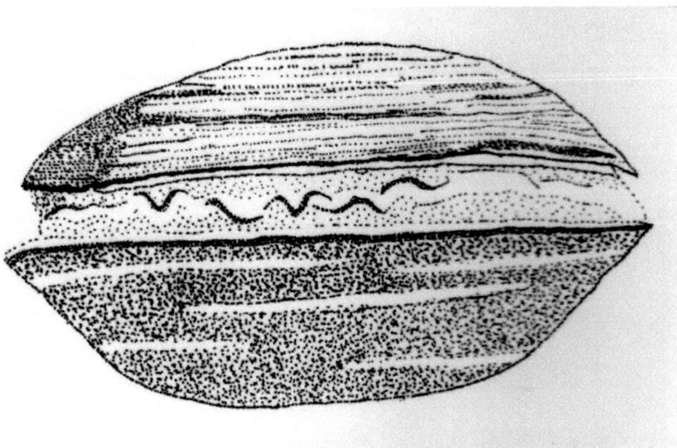

Lost Gaper Clam

Scientific Name: *Tresus pajaroanus*
Order: Veneroida
Family: Mactridae
Size Range: up to 20 cm
Location: West Coast of the United States
Habitat: lower intertidal

About: *Tresus pajaroanus* is a bivalve mollusk commonly known as a lost gaper clam. It is a saltwater clam that lives buried in sand, mud, and gravel substrates in the lower intertidal zone. They are similar to geoducks, but they are smaller and don't bury themselves as far down in the sand. The meat of a similar species of gaper clam can be used in chowder, but sport divers usually ignore them. The Chumash, a Native American people from what is now San Luis Obispo, harvested gaper clams for their shells.

Lucinoma annulata

Ringed Lucina Clam

Scientific Name: *Lucinoma annulata*
Order: Veneroida
Family: Lucinidae
Size Range: up to 10 cm
Habitat: Mainly bays and estuaries

About: Clams are filter feeders so they filter algae out of the water and through their gills; it is carried to their digestive system by tiny hairs-like cilia. Lucina clams do not have a siphon, so, after burying themselves in sand, they make a tunnel to the surface with their foot. Clams are eaten in many different forms, including raw, steamed, made into clam cakes, stuffed, or made into chowder, which is considered the first American soup. Some Native Americans made clamshells into beads that were used as currency. The hole in this fossil was likely made by a predator snail, which might have been what killed it.

Gastropods

Mollusks that have either one or no shell and move around using a ventral muscular foot. This includes snails, slugs, whelks, and limpets.

Crepidula princeps

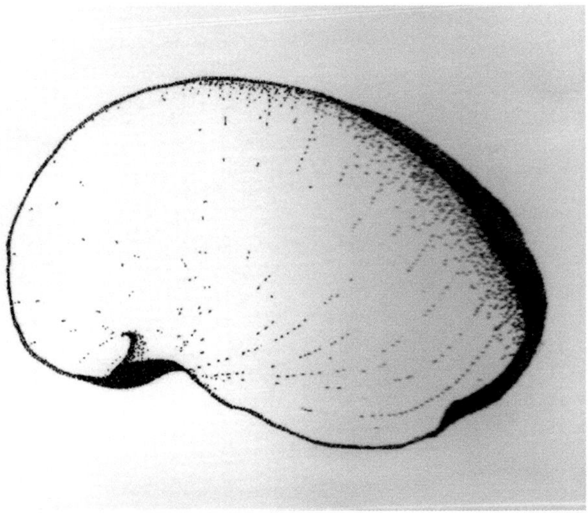

Slipper Shell

Scientific Name: *Crepidula princeps*
Order: Mesogastropoda
Family: Calyptraeidae
Size Range: 2–5 cm
Habitat: Shallow water

About: Slipper limpets are not actually limpets. They are distantly related to limpets, but are more closely related to conches or cowries. They are called slipper limpets because the inside of the shell sort of looks like a slipper. Slipper limpets are also called slipper shells or boat shells, and are a type of snail. Their shells are made of calcium carbonate. Slipper shells are filter feeders so they attach to a rock or another shell and stay there their whole life.

Neverita lewisii

Lorena Paras

Lunatia lewisii

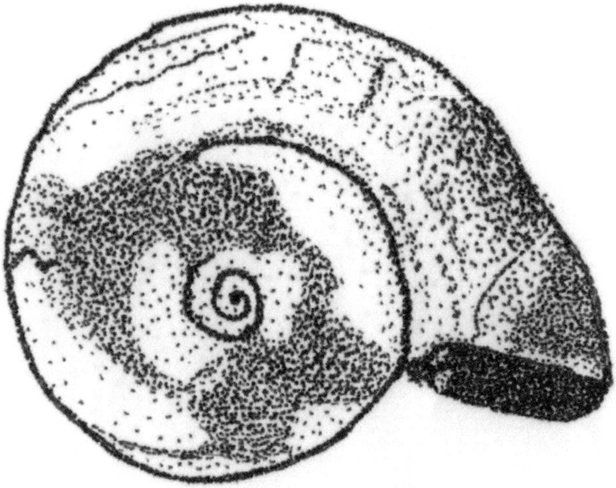

Moon Snail

Order: Mesogastropoda
Family: Naticidae
Size Range: up to 13 cm
Habitat: Sand or mud in intertidal zones or protected bays

About: Snails in the family Naticidae are called moon snails. Moon snails are carnivorous. They eat by enveloping other mollusks (mostly bivalves) with their foot and drilling holes in them to suck out the insides. They will even eat other moon snails. To drill the hole, they use a part of their mouth called a radula, which is kind of like a tongue covered in teeth. Norwegians and other Europeans consider moon snails a delicacy — as do gulls. There are many types of moon snails all over the world including the Arctic and Antarctica.

Nassarius californianus

Nassarius grammatus

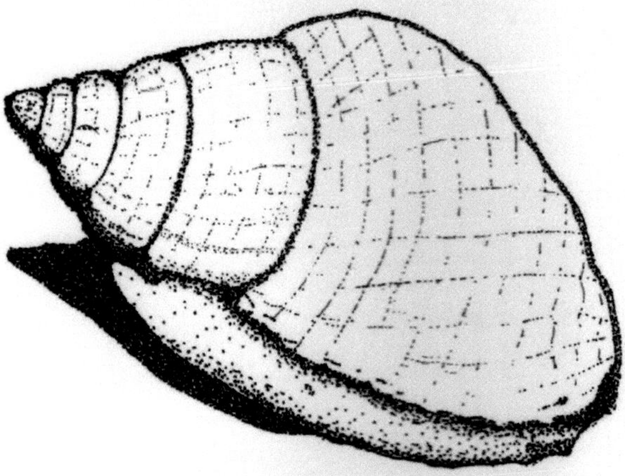

Nassa Mud Snail

Order: Sorbeoconcha
Family: Nasariidae
Location: California
Size Range: up to 3 cm
Habitat: shallow water, sandy or muddy substrates, occasionally in deep water

About: Nassa mud snails are a taxonomic family of small to medium-sized sea snails, marine gastropod mollusks. They are also called dog whelks or basket shells. These snails have rounded shells with a high spire, an oval aperture, and a siphonal notch. They feed on crabs, dead fish, and other carrion. Beads made of the shells of another species from this family are the oldest known jewelry, dated to be over 100,000 years old.

Mammals

A class of animals identified by warm-bloodedness, having hair or fur, and being able to produce milk from mammary glands for their young. Mammals typically give birth to their offspring as opposed to laying eggs.

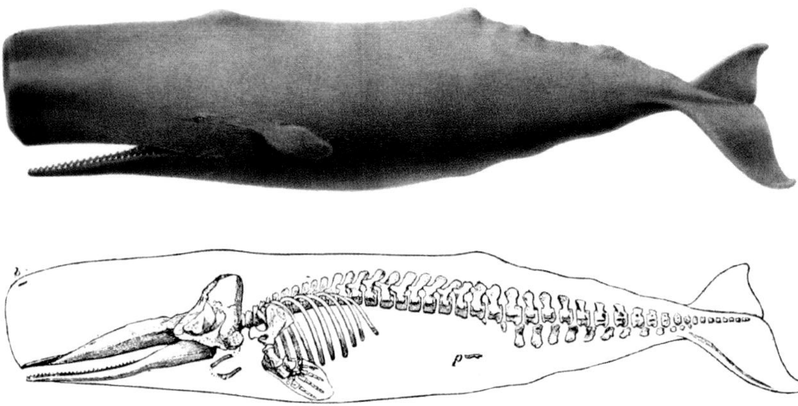

(Sperm Whale skeleton diagram, by Physeter_macrocephalus1.jpg: Citron SpermWhaleLyd3.jpg: Lydekker, Richard derivative work: Totodu74 (Physeter_macrocephalus1.jpg SpermWhaleLyd3.jpg) [Public domain], via Wikimedia Commons)

Lorena Paras

Whale Vertebrae

Whales are marine mammals that live all over the world. They can range in size from 11–98 feet in length. Blue whales are the largest animal to ever exist. Humpback whales can live up to 77 years and Bowhead whales are estimated to have a lifespan of over 200 years.

Most whales have 51 vertebrae, 23 caudal vertebrae, 11 dorsal vertebrae, 10 lombo-sacrel, and 7 cervical. Whales are believed to have evolved from land-dwelling animals to inhabitants of the sea over the course of millions of years.

The design and curvature of a whale's spine resembles a design that appears to have originally been used for running as opposed to swimming.

Lorena Paras

Whale vertebra

Lorena Paras

Whale Vertebra

This appears to be a whale vertebra that is completely embedded in the surrounding rock. This is the most common fossilization process. Millions of years ago, this whale died and its body settled on the sea floor. All of the soft tissues decayed, leaving just the bones. Then water seeped into the spaces within the bones and the minerals dissolved in the water crystallized, which caused the bones to harden along with the encasing rock. This is called permineralization or petrification.

Echinoderms

A marine invertebrate recognizable by radial symmetry in adults — for example sea stars, sea urchins, and sea cucumbers.

Dendraster gibbsii

Sand Dollar

Scientific Name: *Dendraster gibbsii*
Order: Echinarachniidae
Family: Clypeasteroida
Location: Northern Hemisphere
Size Range: typically 5–8 cm
Habitat: sandy bottoms below the low tide level
About: Sand dollars are just the tests of a type of sea urchin. When the sea urchins are alive, they look like small disks covered in colorful fuzz. But when they die, the only part that's left is the skeleton, or test, which we know as a sand dollar. These sea urchins live on the sea floor. To get food, they capture plankton with spikes on their surface, and other spikes and tiny hairs called cilia pass it down to their mouth on the bottom side. When the current gets strong these sea urchins lie flat against the sand or bury themselves. Adults can grow thicker skeletons to fight the currents, and young ones swallow sand to weigh themselves down.

Crustaceans

A subphylum of arthropods that possess a segmented body and an exoskeleton, like crabs, lobsters, shrimp, and barnacles.

Balanus concavus

Acorn Barnacle

Scientific Name: *Balanus concavus*
Order: Sessilia
Family: Balanidae
Size Range: 2–10 cm
Habitat: rocky shores, hard objects

About: Barnacles are sessile arthropods, which means they latch onto one spot and don't move. Their shell grows directly onto the substrate. Barnacles live mostly in shallow waters or the intertidal zone. Barnacles have adapted their shells to compensate for water loss when the tide goes out. They feed by opening up the calcium plates that cover the hole at the top, and feathery legs, called cirri, sift through the water for food. When they are not feeding, their shells are impermeable to water and protect against predators.

About the Author

Lorena Paras is a young author from Northern California with passionate interests in the arts and sciences. Fossil finding has always been a favorite activity, one of many providing depth to her relationship with the Monterey Bay, nearby coastal beaches and sloughs. Several of her interests will be obvious from the pages of this book: photography, drawing, and discovery — particularly the discovery of things that give her a window into the lives of life forms around her. Behind the scenes she enjoys hiking her beautiful coastal beaches, doing art with talented local artists, and chatting up the resident scientist. While this book was created during her 8th grade studies at a Berkeley middle school, at the time of publication Lorena is a junior at Berkeley High School.